Rourke
Educational Media
rourkeeducationalmedia.com

Animals Have Classes Too!

Reptiles

Lisa Colozza Cocca

Before, During, and After Reading Activities

Before Reading: Building Background Knowledge and Academic Vocabulary

"Before Reading" strategies activate prior knowledge and set a purpose for reading. Before reading a book, it is important to tap into what your child or students already know about the topic. This will help them develop their vocabulary and increase their reading comprehension.

Questions and activities to build background knowledge:
1. *Look at the cover of the book. What will this book be about?*
2. *What do you already know about the topic?*
3. *Let's study the Table of Contents. What will you learn about in the book's chapters?*
4. *What would you like to learn about this topic? Do you think you might learn about it from this book? Why or why not?*

Building Academic Vocabulary
Building academic vocabulary is critical to understanding subject content.
Assist your child or students to gain meaning of the following vocabulary words.
Content Area Vocabulary
Read the list. What do these words mean?

- *ancestors*
- *cells*
- *classify*
- *common*
- *extinct*
- *fossils*
- *prey*
- *scales*

During Reading: Writing Component

"During Reading" strategies help to make connections, monitor understanding, generate questions, and stay focused.
1. *While reading, write in your reading journal any questions you have or anything you do not understand.*
2. *After completing each chapter, write a summary of the chapter in your reading journal.*
3. *While reading, make connections with the text and write them in your reading journal.*
   a) Text to Self – What does this remind me of in my life? What were my feelings when I read this?
   b) Text to Text – What does this remind me of in another book I've read? How is this different from other books I've read?
   c) Text to World – What does this remind me of in the real world? Have I heard about this before? (News, current events, school, etc.…)

After Reading: Comprehension and Extension Activity

"After Reading" strategies provide an opportunity to summarize, question, reflect, discuss, and respond to text. After reading the book, work on the following questions with your child or students to check their level of reading comprehension and content mastery.
1. *How do scientists classify reptiles? (Summarize)*
2. *What can you conclude about reptiles within a genus as compared to reptiles within a phylum? (Infer)*
3. *What three things do scientists look at when they sort living things into kingdoms? (Answering Questions)*
4. *If you had to make a new way to classify reptiles, what rules would you make? (Text to Self Connection)*

Extension Activity
Find online information about a reptile family. Draw a family tree. Use the information you find to fill in the branches.

Table of Contents

Let's Classify! 4

Reptile Class 9

Reptile Families 12

Reptile Species 16

Activity 21

Glossary 22

Index 23

Show What You Know 23

Further Reading 23

About the Author 24

iguana

Let's Classify!

Many scientists study living and once-living things. They **classify** or sort these things into groups. They look for ways these things are alike and different.

Animal Groups

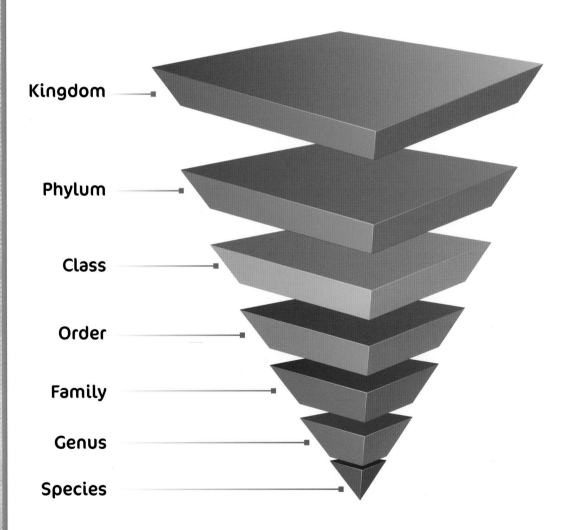

Kingdom

Phylum

Class

Order

Family

Genus

Species

Kingdoms are the biggest groups. Scientists sort things into kingdoms based on three things. The first thing is what kind of **cells** it has. The second is how many cells make up its body. The third is whether or not it makes its own food.

Cytologists are scientists who study cells. They study cells to see how they are made and how they work.

Members of the animal kingdom have bodies made of more than one cell. They cannot make their own food. They eat living or once-living things. Some eat plants. Some eat bugs. Some eat other animals!

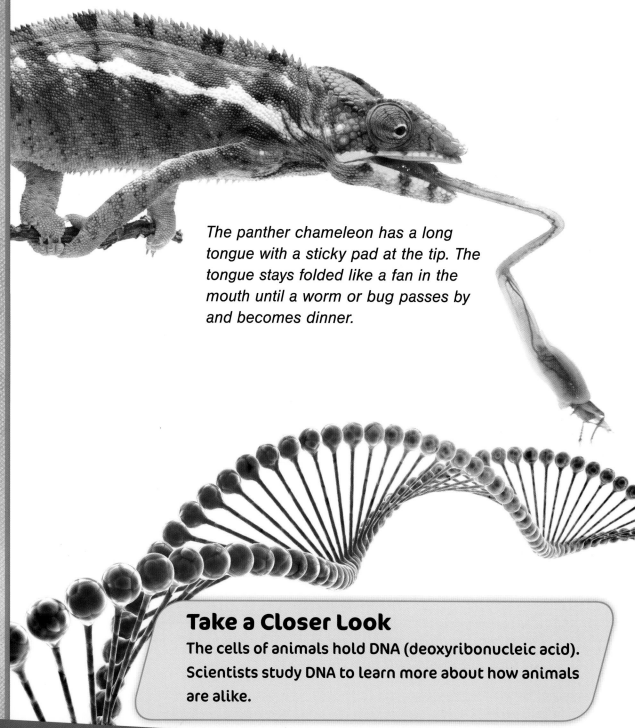

The panther chameleon has a long tongue with a sticky pad at the tip. The tongue stays folded like a fan in the mouth until a worm or bug passes by and becomes dinner.

Take a Closer Look

The cells of animals hold DNA (deoxyribonucleic acid). Scientists study DNA to learn more about how animals are alike.

There are more than a million kinds of animals. Scientists break the kingdom into smaller groups called phyla. Animals with a backbone belong to one group, or phylum. A backbone is a series of connected bones. It runs from the head through the middle of the back.

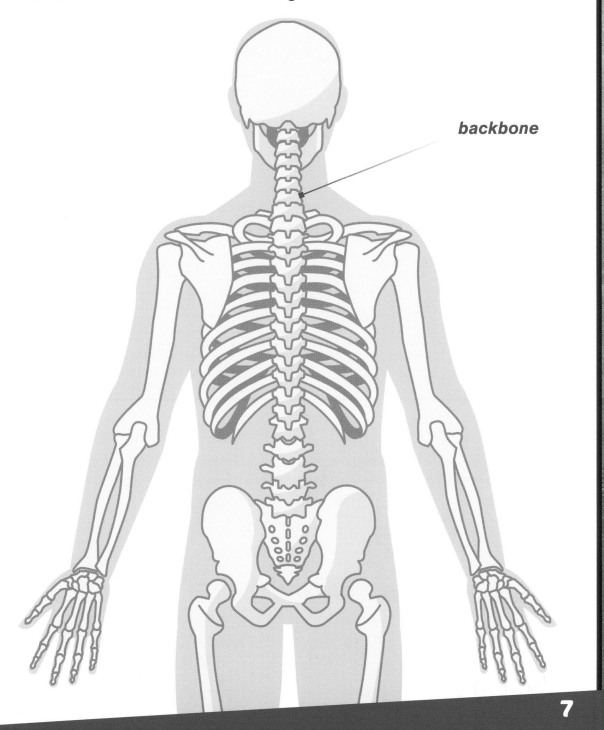

backbone

Many living things have backbones. People have backbones. Elephants have backbones. Turtles have backbones too. People, elephants, and turtles are very different though.

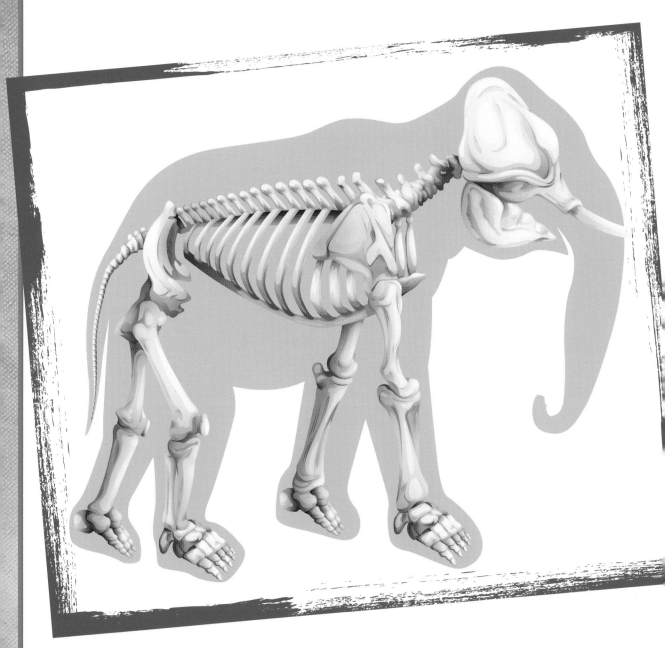

There are more than 300 bones in an elephant's skeleton. The bones make up 16.5 percent of the elephant's total weight.

Reptile Class

Scientists break a phylum into smaller groups called classes. Reptiles make up the class Reptilia. Reptiles are cold-blooded. The bodies of cold-blooded animals copy the temperature of the surrounding air. They warm in the sun and cool in the shade.

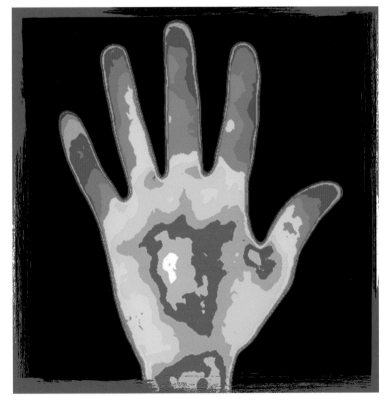

Special cameras are used to take images like this. The cameras take images of the heat given off by the subject.

Body Temperature

People are warm-blooded. Our bodies can make heat to help maintain our body temperature.

All reptiles also have **scales**. Scales come in all sizes. The scales on a dwarf gecko are tiny. The scales on a snake are much bigger. Scales can be soft or hard. Scutes are the biggest kind of scales. They are hard. A turtle shell has scutes on it.

scute

Classes are broken into smaller groups called orders. The reptile class has four main orders. Snakes, lizards, and worm lizards make up the biggest order. Turtles and tortoises make up another order. A third group includes crocodiles, gharials, caimans, and alligators. The smallest order is made up of tuataras.

tortoise

turtle

Tortoise or Turtle?
Tortoises live only on land. Turtles have webbed feet and most live in water all or part of the time.

Reptile Families

Families are the groups that make up an order. Snakes belong to the order Squamata. A ball python is a kind of snake. Its body does not make poison. The snake kills its **prey** by wrapping its body around it and squeezing.

ball python

A copperhead is another kind of snake. Its body makes poison. A copperhead kills by biting and forcing the poison into its prey. Both snakes belong to the same order but are members of different families.

copperhead

We Are Not the Same!

Crocodiles and alligators are both reptiles. They belong to the same order but different families. There are many differences between them.

Scientists break families into genera. Reptiles within a genus are closely related. The sea turtle family is broken into six genera.

Kemp's ridley sea turtle

Kemp's ridley sea turtle and the olive ridley sea turtle belong to the same genus. They look much alike and are about the same size. Both live mainly in the water, but lay their eggs on the beach. One difference is the olive ridley lives in warmer waters than the Kemp's ridley.

olive ridley sea turtle

Reptile Species

A genus is made up of one or more species. A reptile species is a single kind of reptile.

American crocodile

Close Relatives

The American crocodile and the mugger crocodile are two of the 12 species that make up the *Crocodylus* genus.

Each species has a **common** name and a two-part scientific name. The first part tells to which genus the reptile belongs. It always starts with a capital letter. The second part tells the species name. It begins with a lowercase letter. It is the same all around the world.

mugger crocodile

The mugger crocodile has a wide, flat snout like an alligator. Its large fourth tooth proves it is a true crocodile though!

"Komodo dragon" is a common name. Its scientific name is *Varanus komodoensis*. Its common name helps people picture in their minds the big lizard. The scientific name tells us in which genus the species belongs.

Big Enough for Two Names!
Another common name for the Komodo dragon is Komodo monitor. It is the largest and heaviest lizard in the world.

Many species, genera, families, and orders of reptiles are **extinct**. Scientists learned about these early reptiles by studying **fossils**. They learned many of these extinct reptiles roamed Earth millions of years ago.

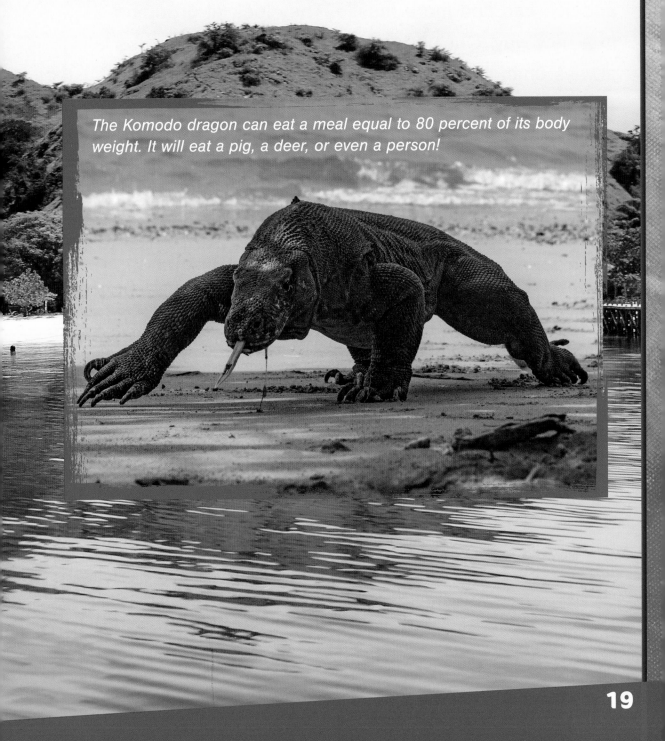

The Komodo dragon can eat a meal equal to 80 percent of its body weight. It will eat a pig, a deer, or even a person!

The groups reptiles are sorted into sometimes change. Scientists once thought a tuatara was a lizard. Then they learned more about the tuatara's **ancestors**.

Scientists are always looking at new ways to classify reptiles. They learn new information. They discover new species and find more fossils. They look at grouping in a new way. Our understanding of the reptile class is always changing and growing.

tuatara

A Living Fossil

Scientists call the tuatara a living fossil. It helps them learn more about reptiles that lived among the dinosaurs.

ACTIVITY

Before and After Pictures

Scientists think all crocodiles share an ancestor: the aetosaur. This reptile roamed Earth hundreds of millions of years ago. It had a body shaped like a crocodile. It had a head shaped like a bird's. Its nose was shaped like the snout of a pig!

Supplies

paper

pencil

crayons or markers

internet with adult's permission

Directions

1. Look up aetosaurs online to see what they may have looked like. Draw one.
2. Draw a picture of a crocodile.
3. Compare your pictures. What are the differences? Then think about how the crocodile might look millions of years in the future. What changes might happen? Draw a picture.

Glossary

ancestors (AN-ses-turz): family members that lived long ago

cells (SELz): the small basic units that make up a living thing

classify (KLAS-uh-fye): sort into groups of alike things

common (KAH-muhn): used by many people

extinct (ik-STINGKT): no longer found living

fossils (FAH-suhlz): traces or prints in a rock or earth of the remains of something that lived long ago

prey (PRAY): an animal that is hunted by another for food

scales (SKALEz): thin, overlapping pieces of skin that cover a reptile's body

Index

backbone(s) 7, 8

ball python 12

cold-blooded 9

copperhead 13

Komodo dragon 18, 19

scientific name 17, 18

scute(s) 10

tuatara 20

turtle(s) 8, 10, 11, 14, 15

Show What You Know

1. Why do scientists break living things into groups?
2. What do all members of the animal kingdom have in common?
3. Which reptiles are more closely related: all members of the same family or all members of the same genus?
4. How can you know if a living thing is a reptile?
5. Why are fossils important to scientists?

Further Reading

National Geographic Kids, *National Geographic Kids Reptiles and Amphibians Sticker Activity Book*, National Geographic Children's Books, 2017.

Riehecky, Janet, *Reptiles*, Capstone Press, 2017.

Wilsdon, Christina, *Ultimate Reptileopedia: The Most Complete Reptile Reference Ever*, National Geographic Children's Books, 2015.

About the Author

Lisa Colozza Cocca has enjoyed reading and learning new things for as long as she can remember. She lives in New Jersey by the coast. You can learn more about Lisa and her work at www.lisacolozzacocca.com.

Meet The Author!
www.meetREMauthors.com

© 2019 Rourke Educational Media

All rights reserved. No part of this book may be reproduced or utilized in any form or by any means, electronic or mechanical including photocopying, recording, or by any information storage and retrieval system without permission in writing from the publisher.

www.rourkeeducationalmedia.com

PHOTO CREDITS: Cover and Title Pg ©amwu; Border ©enjoynz; Pg 3 ©HuyThoai; Pg 4 ©lvcandy; Pg 5 ©Zinkevych; Pg 6 ©Jezperklauzen, ©stanley45; Pg 7 ©elenab; Pg 8 ©blueringmedia; Pg 9 ©AnitaVDB; Pg 10 ©Nerthuz; Pg 11 ©Dzmitrock87, ©AlizadaStudios; Pg 12 ©MirekKijewski; Pg 13 ©makasana; Pg 14 ©Jereme Phillips, ©USFWS.; Pg 15 ©sdbower, ©LagunaticPhoto; Pg 16 ©Patrick Gijsbers; Pg 17 ©sabirmallick, ©YinYang; Pg 18 ©ANDREYGUDKOV; Pg 19 ©AlfinTofler, ©guenterguni; Pg 20 ©Beautifulblossom; Pg 21©Ana Nhtepheta; Pg 22 ©PetlinDmitry

Edited by: Keli Sipperley
Cover and interior design by: Kathy Walsh

Library of Congress PCN Data

Reptiles / Lisa Colozza Cocca
(Animals Have Classes Too!)
 ISBN 978-1-64369-062-9 (hard cover)
 ISBN 978-1-64369-077-3 (soft cover)
 ISBN 978-1-64369-209-8 (e-Book)
Library of Congress Control Number: 2018955963

Rourke Educational Media
Printed in the United States of America,
North Mankato, Minnesota